BRITISH GOATS

Tiffany Francis

SHIRE PUBLICATIONS

Bloomsbury Publishing Plc

PO Box 883, Oxford, OX1 9PL, UK

1385 Broadway, 5th Floor, New York, NY 10018, USA

E-mail: shire@bloomsbury.com

www.shirebooks.co.uk

SHIRE is a trademark of Osprey Publishing Ltd

First published in Great Britain in 2019

A catalogue record for this book is available from the British Library.

ISBN: PB 978 1 78442 360 5

      eBook 978 1 78442 359 9

      ePDF 978 1 78442 357 5

      XML 978 1 78442 358 2

19 20 21 22 23   10 9 8 7 6 5 4 3 2 1

Typeset by PDQ Digital Media Solutions, Bungay, UK

Printed and bound in India by Replika Press Private Ltd.

Shire Publications supports the Woodland Trust, the UK's leading woodland conservation charity.

COVER IMAGE
Front cover: Male Bagot goat (Alamy). Back cover: *Signs of the Zodiac – Capricorn*, by Hans Thoma, c.1909 (National Museum in Warsaw/Public Domain).

TITLE PAGE IMAGE
British Alpine goat.

CONTENTS PAGE IMAGE
A kid gets comfortable in the food tub.

ACKNOWLEDGEMENTS
Images are acknowledged as follows:

Culture Club/Getty Images, page 17; Cyrus McCrimmon /Getty Images, page 36; DEA/C.Sappa/ Getty Images, page 32; DEA/V.Gianella/Getty Images, page 51 (bottom); Emilie Chaix/Getty Images, page 10 (bottom); English Heritage/Heritage Images/ Getty Images, page 11; Farm Images/Getty Images, page 35; FLPA/David Hosking/Getty Images, page 42; George W. Hales/Stringer/Getty Images, page 31; Greg Wood/AFP/Getty Images, page 30; Historica Graphica Collection/Heritage Images/Getty Images, page 18; Hulton Deutsch/Getty Images, page 38; ilbusca/Getty Images, page 10 (top); iStock, pages 1, 22 (left), 27 (bottom), 43 (left), 43 (right), 45, 46 (top), 46 (bottom), 51 (top), 54 (bottom); John Paul Getty Museum, page 12 (top); Martin Bureau/AFP/ Getty Images, page 54 (top); Mary Evans Picture Library, pages 13 and 14; Nature Picture Library/ Getty Images, page 12 (bottom); Neil Setchfield/ Getty Images, page 44; Niall Benvie/Nature Picture Library/Getty Images, page 39; Paul Nicholls/ Barcroft Media via Getty Images, page 56; Peter Muller/Getty Images, page 6; Pixabay, pages 20, 21, 22 (right), 24, 25, 26 (top), 26 (bottom), 27 (top), 28 (bottom), 29, 37; Private collection/Wikimedia Commons, page 19; Russ Carmack/Tacoma News Tribune/ MCT via Getty Images, page 50; Saqib Majeed/ SOPA Images/LightRocket via Getty Images, page 55; Steve Austin/Getty Images, page 52; Sylvain Cordier/ Gamma-Rapho via Getty Images, page 28 (top); The Washington Post/Getty Images, page 49 (top right); Universal History Archive/UIG via Getty Images, page 8; Wikus De Wet/Getty Images, page 33.

All other images are from the author's own collection.

CONTENTS

INTRODUCTION

Two yellowhammers are calling from the hedgerow, and beyond them the endless warmth of summer pours over the quiet settlement of Butser Ancient Farm in Hampshire. I watch the swallows sweep in and out of our Iron Age roundhouse, the roof of which has been their favourite nesting spot since the house was built more than ten years ago. To my left, one member of our English goat herd, Áine, is pushing her forehead gently into my shoulder while I scratch the little space between her horns. Behind her the other members of the herd, Sorrel, Bella and Yarrow, are all squeezed onto the same log in an attempt to stand taller than the others. It's crowded, and my hair is regularly chewed, but this is the real joy of goatkeeping.

My passion for goats began when I started working at Butser Ancient Farm, an archaeological research site in the South Downs, partly run by my brother-in-law Simon. In the months before, I'd been living in London while studying for my Master's degree in English Literature, and when I finally recognised how incompatible I was with city living, I moved back home to Hampshire and started working on the farm. It is a strange and beautiful place in the heart of the Downs – a living, breathing monument to ancient Britain, complete with Iron Age roundhouses, a Roman villa, Saxon hall and a collection of rare breed and traditional livestock.

When I first joined the farm I was already vegetarian, and while I gave equal love and care to all the animals we kept

Tiffany with the goats at Butser Ancient Farm.

there, I could never form too strong a bond with them because I knew most would end up in the freezer. The only animals we didn't send to slaughter were our herd of English goats, which we kept for their historical significance, low maintenance costs, and because the visitors loved them. Knowing they were able to live out their lives in full, I allowed myself to nurture a real friendship with the herd, and over the next few years I not only learnt about animal husbandry and livestock health, but also discovered how intelligent, mischievous and independent

Goats are highly intelligent and love human company.

they were. I learnt how to milk them and make hard, soft and blue cheeses, took them to win rosettes at agricultural shows, and even helped them give birth. I brushed their coats during the moulting season and fed them banana skins on warm afternoons, and over time I grew to understand their changeable social hierarchies.

While commonly mistaken as stupid, goats are confident, intuitive and compassionate animals, with one recent study at Queen Mary University in London suggesting they are as intelligent and loving as dogs. At the farm we knew for a fact they were only remaining in their paddock because they chose not to escape, a point proven when there was a thunderstorm one night, and the next morning we found goat droppings leading all the way up the staircase to the staff room. We later found the herd huddled in their shack, faux innocence on every face.

Goats are full of surprises. I like filming wildlife on my trail camera, capturing badgers and foxes on their nocturnal excursions across the farm. One morning I flicked through my footage to find that I'd accidentally pointed the lens too close to the goat paddock. We had assumed the goats went to sleep after dark, but the footage showed that, instead of retreating to the safety of their shack, they roamed around freely through the night. Every clip of motion-triggered footage showed a pair of eyes drifting mysteriously across the screen, searching for stray hazel leaves under cover of darkness.

All animals, both wild and domesticated, are fascinating creatures, but goats will always have a special place in my life. In researching this book, I discovered that not only do I fall under the Capricorn (sea goat) star sign in the modern zodiac, but I was also born in the Chinese year of the goat, so I can't help feeling the universe may have gently nudged me towards goats and goatkeeping. I hope this book will enable more people to understand the brilliance of these animals, and help to illuminate the rich heritage of domesticated goats throughout the British Isles.

A SURVIVAL STORY

Until the late eighteenth century, the only domesticated goat breed widely established in this country was the British primitive goat, brought over to Britain towards the end of the Stone Age around 5,000 years ago. Originally derived from the bezoar (*Capra aegagrus*) native to the Middle East, the British primitives were some of the first animals to be domesticated during the Neolithic period, together with sheep and cattle. When the ancient Britons stopped moving around in temporary settlements, following herds of animals across their habitats, they began instead to build permanent communities in one place and to domesticate wild animals as livestock.

During the Bronze and Iron Ages, both sheep and goats were kept for their meat, skin, hair, wool, hoof, horn, fat, milk, and for transport. According to the scornful Greek geographer Strabo, however, the earliest British farmers were not skilled dairymen, 'without even sense enough to make cheese, though milk they have in plenty.' When the Romans arrived, sheep became the more favourable livestock due to the usefulness of their wool, except at some Roman temple sites like Uley in Gloucestershire. Here, archaeologists have discovered the remains of goats butchered in curious ways, suggesting they might have been condemned to the sacrificial altar.

By the eighth century the Vikings had invaded, bringing with them a passion for the dairy industry. According to legend, the Vikings were among the first to refine cream

OPPOSITE
Illustration from the Goat Show at the Crystal Palace, 1883.

Antique illustration of a bezoar.

into butter and trade with it as a commodity, and when they settled in Britain they transformed the quiet Celtic farmsteads they found into small, bustling towns. The goats adapted to living amongst people in more cramped conditions, and it wasn't until the Normans arrived in 1066 that they were able

Petroglyph depicting a hunting scene from Central Asia, second century BC.

to return to their natural free-roaming behaviour in herds. Under the medieval feudal system, goats were released into the manor grounds where they were able to explore woodland and pasture, feeding on fresh scrub and wasteland avoided by other livestock, while still enjoying human care and companionship at the cottage door. Prized for their resilience and delicious milk, these were the glory days for the British primitive goat – but sadly, they were not to last.

Prior to the Enclosure Acts of the late eighteenth century, it was common practice for herds to be taken to higher ground for the summer and brought back again for winter, and it was during this process that some goats escaped and formed their own feral herds in the hills. Historians now believe that if it weren't for these hardy escapees, the British primitive goat may have disappeared altogether. By the end

Illustration of an Iron Age roundhouse.

A thirteenth-century depiction of goats from an English bestiary.

of the Middle Ages, sheep had become the favoured stock of upland farmers and goats had fallen out of fashion in Britain, although they remained valued throughout the rest of Europe. It is unclear exactly why this happened, but theories suggest their dwindling popularity was due to snobbery across the social spectrum.

The Enclosure Acts were a series of legal processes in England that consolidated hundreds of small landholdings into larger farms, transforming what was once common land for communal use into privately owned land. The Acts

British primitive goat by the sea.

Shepherdess
with goats
and sheep, by
George Romney
(1734–1802).

were implemented over many decades so that by the start of the First World War, 95 per cent of the British landscape had been enclosed. Goats had already fallen in popularity by the time Enclosure set in, but the Acts triggered a top-down cascade when the aristocracy began to reject goats from being kept on their land, fearing the animals might damage their grounds and ornamental gardens. This caused independent farmers to remove them from their land, fearing they would damage the hedgerows ignored by the sheep and cattle. Despite the utility and resilience of goats, in the end even the lower classes abandoned them, perhaps with the mindset that if goats weren't good enough for their superiors, they weren't good enough for them. Within a few centuries the British primitive goat, although just as productive, useful and hardy as it had ever been, had become a rare and vilified creature, so that by the nineteenth century, country magazines were commenting that 'few could nowadays venture to harbour such mischievous and dangerous animals.'

Angela
Burdett-Coutts
(1814–1906).

In 1875, a letter appeared in *The Times* that would mark the beginning of what has become known as the Victorian goat revival. The philanthropist Baroness Angela Burdett-Coutts, who gave £3 million to good causes over her lifetime, wrote to the newspaper about the 'milk' trains that carried milk from poor rural towns to feed the inner cities. She had noticed that the poorest children who lived in the countryside received none of this milk and were malnourished, and suggested they keep goats to feed themselves. The popularity of goatkeeping had quietly started to rise again, and she had heard of their resilience and nutritious milk; in fact her own goat, 'Polly', became the first entry in the herd book for the British Goat Society.

That summer, the first agricultural show for goats was held at the Crystal Palace, a cast-iron and plate-glass structure in London originally built to house the Great Exhibition in 1851. Four years after the show, the British Goat Society was founded with Baroness Burdett-Coutts as the first Patron. Goatkeeping regained its fashionable status and became a more middle-class affair, and the prejudice against goats slowly began to wane. By this point, trade routes had opened up across the globe, and rather than breeding solely from the remaining British primitive stock, goatkeepers crossbred with more exotic breeds brought in from Europe, India, Africa and the Middle East. These included the Toggenburg and Saanen from Switzerland, and lop-eared goats from India, which were crossbred to become the Anglo Nubian.

While goatkeeping continues to rise in popularity, the British primitive goat that once roamed Stone Age settlements, feudal farms and Viking towns is now considered a rare breed with around 5,000 individuals remaining across forty-five populations, although some researchers claim only around 1,500 of these are purebred. Today, feral herds still exist in the Burren in the west of Ireland; Snowdonia in Wales; Lynton in Devon; the Scottish isles of Jura, Mull and Rum; northern Scotland; Dumfries and Galloway; Northumberland; and the Isle of Wight. Small, managed herds are also kept for conservation grazing, like those at Cheddar Gorge and Windsor Great Park, and a number are kept as visitor attractions in zoos and wildlife parks.

There are now around seventeen breeds of goat established in Britain, including the descendants of a small herd of Bagot goats thought to have been brought over during the Crusades in the fourteenth century. Most have their own breeding society, which helps to keep standards high and prevent interbreeding, and goats have now become a prized element of agricultural shows across the country. In modern Britain, goats are kept for milk and cheesemaking, clearing land and controlling shrubs, organic meat, cashmere wool, rural education, manure, as companions to other animals, or simply as endlessly fascinating pets.

Feral goats roaming through Cheddar Gorge.

CULTURE, MYTH AND FOLKLORE

IN THE WESTERN world, the goat is an ancient symbol of fertility, vitality, sexuality and boundless energy, perhaps due to the insatiable behaviour of billies at mating time. In fact, the aroma of uncastrated males is so powerful that the mineral bromine is named after the Greek word *brómos*, meaning 'the stench of he-goats'. A common superstition from the Middle Ages claimed that goats whispered lewd words into the ears of saints, and their fondness for escaping associated them with disobedience and sin, although in reality, this was probably down to biology rather than immorality.

When the Greek slave and storyteller Aesop recorded his *Fables* over two thousand years ago, the goat was usually portrayed as stupid, lazy or arrogant. In one story, a goat hides from its hunters behind a tendril in a vineyard. Once the hunters pass, he begins to nibble at the leaves and the vine asks him, 'Why do you injure me without a cause?' He continues to eat greedily, but the noise he makes soon brings the hunters back. They slaughter him and the vine is avenged. One of the better-known stories tells of a fox who has fallen into a well. Struggling to find a way out, the fox sees a goat approach the well and call down to ask if the water is plentiful. The fox assures him the water is sweet and tasty, and suggests the goat jump down to join him in the well and quench his thirst. The goat immediately leaps into the well, whereupon the fox hops onto his back and horns to climb out, laughing to the goat as he leaves: 'If you had half

as many brains as you have beard, you would have looked before you leapt!'

The Eastern traditions are more forgiving of our bearded friends. In Chinese astrology, goats are depicted as peace-loving, trustworthy and kind, and those born in the Chinese year of the goat are said to be shy, introverted and creative. In ancient Judaism, goats played a central role in the celebration of Yom Kippur, the holiest day of the Jewish year. Followers were encouraged to atone and repent for their sins, and as part of the ceremony two goats were chosen by lottery. One of these was offered as a blood sacrifice, and the other was symbolically burdened with the sins of its people and cast out into the desert to carry them away. This is the origin of the word 'scapegoat'.

In C.S. Lewis' fantasy series *The Chronicles of Narnia*, Lucy Pevensie meets a faun called Mr Tumnus. This half-human-half-goat is thought to have inspired the entire series,

Illustration of *The Goat and the Vine* by Arthur Rackham.

Illustration from *The Wind in the Willows* by Paul Branson, 1913.

when the author claimed an image came to his mind of a faun carrying an umbrella and parcels through a snowy wood. With his pointed beard, short horns, cloven hooves and a 'strange but pleasant little face', Mr Tumnus holds none of the sinful qualities often attributed to goats, going on to become a hero of the series and a loyal friend to the four Pevensie children.

A familiar figure in both Roman and Greek mythology, the goat-man appears in another children's classic, first published in 1908. Kenneth Grahame's novel *The Wind in the Willows* follows Mole, Rat, Badger and Toad on their adventures in a pastoral England. In the seventh chapter, entitled 'Piper at the Gates of Dawn', Rat and Mole meet the Greek deity Pan, who helps them recover the Otter's lost son Portly. God of the wild, shepherds, mountains and rustic music, in Greek mythology Pan has the hindquarters, legs and horns of a goat and lives in the unspoiled wilderness known as Arcadia. He has since become an icon of the golden age of pre-civilisation, when mankind was more in touch with its wild, primitive origins, and has influenced poets, writers, artists and musicians (including Pink Floyd). His name was even referenced in J.M. Barrie's *Peter Pan*, whose titular character is both sweet and selfish, part human and part primitive child.

Another piece of caprine mythology is used to explain the constellation of Capricornus, the smallest constellation in the zodiac. The exact origins of the Capricorn myth are unclear, but one theory adheres it to the Greek tale of Pricus, an immortal sea goat who could manipulate time. Favoured by the Greek gods, he enjoyed a happy life by the seashore with his many children, but when each of his young found themselves on dry land they turned back into normal goats, losing their intelligence and ability to speak. In an attempt to stop his children finding their way back to land, Pricus turned back time again and again, but in the end resigned himself to loneliness and watched each of his children leave him. Not wanting to be the only sea goat left, he asked the time deity Chronos to let him die, but due to Pricus' immortality, he was refused. Instead, Chronos placed him in the heavens to spend eternity as the constellation Capricorn.

Pan and Psyche by Gustav Klimt, 1892.

In Old Norse culture, the hammer-wielding god Thor, associated with thunder, lightning, oak trees, strength and fertility, kept two goats as steeds to pull his chariot. Their names were Tanngrisnir and Tanngnjóstr, meaning 'teeth-barer' and 'teeth-grinder'. In thirteenth-century Old Norse texts, the story goes that at the end of each day's travel, Thor slaughters his goats and feasts on their flesh, resurrecting them afterwards with his hammer so they can continue to serve him. One

night, Thor and the shapeshifting god Loki stop at the home of a peasant family, looking for a night's lodging. As usual, Thor slaughters his goats, skins them and places them in the pot, inviting the peasant family to share their feast. At the end of the meal, Thor places the skins on the opposite side of the fire and instructs the peasants to throw their goat bones onto the skins. Instead, the peasant's son Þjálfi uses a knife to slit one of the bones open and eat the marrow, before throwing it on the pile with the rest. The next morning Thor wakes and resurrects his goats, but notices that one of them is now lame in the hind leg. Enraged, he accuses the peasant family of mistreating the bones and disobeying his instructions, but on seeing how terrified the family are, he calms himself and accepts as settlement the service of the two children Þjálfi and Röskva. The children become his servants for eternity and the goats are left behind.

Statue of Peter Pan in Kensington Gardens, London.

According to the contemporary pagan movement known as Wicca, another deity exists in the form of the Horned God, associated with nature, sexuality and wilderness, and thought to emphasise the union of the divine and the animal. One theory suggests he carries the souls of the dead to the underworld, while another claims this was the original inspiration for the Krampus from alpine culture, a horned figure who punishes naughty children at Christmas time. Described as having black hair, cloven hooves, horns, fangs and a lolling tongue, the Krampus' arms are chained and often adorned with bells. He carries bundles of birch branches known as *Ruten*, which he uses to swat children before lifting them into the basket on his back and carrying them off to drown, be eaten or led into Hell.

In Ancient Egypt, the horned god Banebdjedet was worshipped as the god of the source of the Nile. Each year when the Nile flooded, it brought silt, clay and water, which he combined at a

St Mark's Clock, Venice, depicting the zodiacal constellations.

potter's wheel to create the bodies of human children and place them in their mothers' uteruses. It is Banebdjedet who is thought to have inspired the legend of the Sabbatic Goat, Baphomet, a modern-day symbol for satanic worship, rebellion and evil. In reality, the story of Baphomet is not one of satanic worship, but a strange and fragmented tale originating from the Crusades of the thirteenth and fourteenth centuries.

The earliest recorded use of the name Baphomet comes from a letter written by a French crusader in 1098, who described their enemies in the Holy Land 'calling upon Baphomet' prior to battle. Today, it is commonly accepted that Baphomet referred to Mohammed, the Prophet of Islam, the worship of whom was considered idolatry in Christian Europe. Two hundred years later, King Philip IV of France ordered the arrest of the Knights Templar, a Catholic military order to which he had become heavily indebted. To clear the debt, he accused them of heresy amid rumours the name Baphomet had been incorporated into their religious practice, perhaps due to their time mingling with Islamic culture in the Middle East. To bring an end to their ensuing torture, the

Templars confessed to worshipping a goat-headed idol and the order was disbanded by the Pope.

Over the next five hundred years, the Templars' strange confessions were enriched and diluted, but it wasn't until 1818 that the name Baphomet re-emerged in an essay published in Vienna, claiming the Knights Templar were indeed idol worshippers. In 1854, the French magician Eliphas Lévi re-imagined Baphomet into a figure he named the Sabbatic Goat, inspired by the Egyptian deity Banebdjedet and the ancient god Pan. Today, Baphomet has become one of the most recognisable occult images in the world, appearing in satanic cults, religious propaganda and even heavy metal music, but while many associate the figure with diabolical tendencies, for others, his connection with the tortured Templars has simply made him a symbol of freedom and independence.

ANATOMY AND BEHAVIOUR

Anatomically, goats have evolved to be resilient, hardy and adaptable. They are sure-footed animals, with a rough pad on the bottom of their two-toed hooves, enabling them to climb and grip onto extreme terrains. They use their feet for the majority of personal grooming, scratching the neck and head with their rear toes but licking the rest of their bodies to keep clean. Like horses, hippopotami and other ungulates, goats also have rectangular pupils which allow them to see a 280° panorama around their bodies. Each eye is placed on opposite sides of their heads, which widens their field of vision to almost 360°, allowing them to spot danger quickly and make their escape.

Being ruminants, goats have four stomachs, and their digestive process is specialised from top to tail. They have no upper incisor or canine teeth, relying instead on lower incisors, lips, tongue and a dental plate to take food into their mouth. The food passes first into the rumen, the largest of the four stomachs, which contains microorganisms and enzymes to break down any fibre. The cellulose within this fibre is converted into fatty acids, which are then absorbed through the rumen wall and provide up to 80 per cent of the animal's total energy requirements. Digestion and fermentation continues as the food passes into the reticulum, followed by the omasum, where water is removed from the food. Lastly, it enters the abomasum, where hydrochloric acid and digestive enzymes break down the remaining food

Goats have rectangular pupils to enhance their vision.

particles before they enter the small intestine and make their way out of the body.

In the wild, goat herds are led by a dominant female and dominant male. Compared with sheep, goats are slightly more aggressive and inquisitive in their herds, and tend to demonstrate their dominance more. To do this, they lower their heads and point their horns at each other, locking them together to determine which animal is more powerful. When new goats are added to a herd, fights and squabbles will usually break out while the hierarchy is reshuffled. Dominant males are usually chosen for their horns and body size, and will mate with the females when they come into season. The female, known as the 'queen', leads the way when the herd is foraging for shrubs and greenery, as well as securing the most comfortable sleeping place, but she will also protect the rest of the herd from approaching predators. Like the male, she will usually maintain her dominance until she dies or another female challenges her and takes her place, and any kids born from a dominant female will inherit her elevated position in the pecking order.

Female kids will come into season during their first autumn, but in the UK most pedigree breeders wait until they are around 18 months old before mating them. Male kids can become sexually active at only a few weeks old, which means they must be kept separate from about six weeks if they are not being castrated. Once a female comes into season, she will be ready to mate for as little as a few hours up to four days or more, and if unmated she will return to season every twenty-one days. Signs that a doe is in season include a wagging of the tail, noisy bleating and a wetness around the back end. Once mated, the gestation period for goats is around 150 days.

One of the reasons goats are so well adapted to different conditions and climates is their hair. Most have a soft, downy undercoat covered by a coarse outer coat in various colours and lengths. The fibres of breeds like the Angora goat are prized for their fine, soft quality, and can be used to make mohair and cashmere products and clothing.

A goat cares for her kids.

Balls of mohair
wool from the
Angora goat.

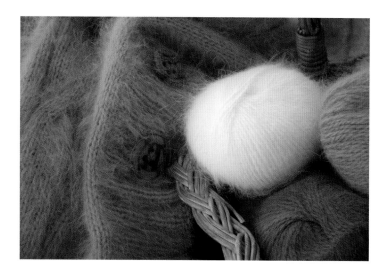

A dairy doe can convert a wide range of vegetation into a rich, nutritious milk, utilising a larger variety of plants that may be inedible or inaccessible to sheep or cattle. Due to their

Saanen goat
eating nettles.

Milking a goat.

hardiness and resilience, they can thrive in a wider range of climates than other livestock, and when given a healthy diet of quality hay, adequate pasture, grain and water, a dairy goat can provide her owner with a tenth of her body weight in milk each day, depending on her breed. This milk can be consumed raw or pasteurised, churned into butter, cream and cheese, or used in cosmetics like soap and moisturiser.

Goats are an excellent source of meat, known around the world as chevon, cabrito or capretto. Chevon is thought to have a higher protein content than beef and lamb, with a lean flavour and a lower fat content than beef or pork. Goats are also valued for their skins, particularly in the United States and Asia where thousands of dry goat skins are used to make a tough form of leather.

Goat meat satay sticks.

Boy with a goatskin bag on the island of Socotra.

Goats cosy in their stable.

Gardeners and farmers can also look to their goats for a reliable supply of manure, with the average goat producing over a ton each year. Unlike horses and cows, goat faeces appear

in pellet form and can be easily collected and distributed, and as a fertiliser goat manure is an excellent source of nitrogen, phosphate, potash and other vital minerals.

Goats are sociable, charismatic animals that respond well to human attention, but they are not solitary and need other animals around them for companionship. Some smallholders even keep sheep or Shetland ponies in with their goats to provide a sense of 'herdship'. Tethering is usually avoided at all costs, as even when carried out correctly it is time consuming and restrictive to the goat's natural behaviour. If poorly managed, it can lead to lack of water and grazing, or exposure to stray dogs, cruel children, heavy rain and hot sun.

A carefree goat relaxes in her paddock.

BRITISH GOAT BREEDS

ANGLO NUBIAN

With a long, deep body and upright stance, the distinctive feature of this breed is its pronounced 'Roman' nose, and long, drooping ears. The short, silky coat comes in a number of colour variations. Although stereotyped as vocal, the breed is quiet when provided with food, water and shelter, and just like a human infant will only make a noise to alert you to their needs.

OPPOSITE
Anglo Nubian goat.

BREED ORIGINS

Anglo Nubians are crossbred between native British goats and a mixed population of lop-eared goats imported from India, the Middle East and North Africa. Their durability means they can live in extreme climates, and have since been exported to more than sixty countries around the world.

Anglo Nubians at the Cambridge Show, 1950s.

PRODUCE

Meat, milk, yoghurt and cheesemaking. Although milk yields can be lower than in other breeds, this is one of the best breeds for butterfat production, with an average of 4.6 per cent or more butterfat content. With a longer breeding season than that of other goats they are also able to produce milk all year round.

ANGORA

The breed most likely to be confused with a sheep, although they are not directly related. The Angora should have a sturdy body and strong legs to support the heavy fleece on its back, which can weigh as much as 6kg. The fleece grows into ringlets with a spiral twist, and goats are sheared twice a year – once in midwinter and again in late summer.

Angora wool.

BREED ORIGINS

Thought to have descended from the markhor (*Capra falconeri*) of central Asia, the first Angora goats in Europe were brought over by Charles V in the sixteenth century, although they were not successfully established until much later.

PRODUCE

The production of fine, luxurious fibre known as mohair, not to be confused with the wool of the Angora rabbit. Mohair is sometimes referred to as the 'diamond fibre' due to its natural strength and lustre. It can be easily dyed to produce bright, rich colours, and is often blended with other natural fibres to produce yarn and other textiles.

BAGOT

An ancient native breed that was once considered 'critically endangered' by the Rare Breeds Survival Trust. The Bagot is medium sized with a long coat, horned head and a nervous character that is common among native breeds of livestock. The ideal colour pattern is an entirely black face from nose to shoulder, followed by a white body behind the shoulder line, although there are variations.

OPPOSITE
Angora goat.

A Bagot goat roams across Snowdon.

BREED ORIGINS

Bagot goats were introduced to England at Blithfield Hall (Staffordshire) in the 1380s, when they are thought to have been brought back by returning Crusaders. King Richard II offered a herd to John Bagot of Blithfield to thank him for the excellent hunting enjoyed in his grounds. Their ancestry may be traced back to the goats of the Rhône valley in southern France.

A Bagot goat at Cheddar Gorge.

PRODUCE

The Bagot has little commercial purpose due to its size and low yields, although the milk it does produce is of a high quality. They are most useful as livestock for conservation grazing in nature reserves where their browsing activity encourages a diverse range of plants beneficial to other wildlife.

BOER

A short-legged, stocky goat with a chocolate-coloured head and neck and a white, short-coated body. They are naturally horned but have such a docile temperament that some owners do not bother to disbud them (remove their horns). Being a fast-growing animal, a mature Boer buck can weigh over 100kg.

BREED ORIGINS

The Boer was developed specifically for meat production in South Africa in the early 1900s. They may have been bred from the indigenous South African goats of the Namaqua,

Boer goat.

A Boer takes a nibble.

San and Fooku tribes, with the possible addition of Indian or European bloodlines. Their name is derived from the Afrikaans word *boer*, meaning farmer.

PRODUCE

In Britain, the Boer is most commonly used as a terminal sire, which means it is used to breed kids for slaughter rather than replacing the flock. They were originally bred to produce lean meat with low cholesterol. Boers can also be used as dairy goats, and due to their gentle nature they make pleasant family pets.

BRITISH

A British goat is one that is registered with the British Goat Society but is not eligible to fall under a particular breed section. This can happen if a goat is crossed with another breed and no longer qualifies as pedigree, or if a goat is registered without full details of its ancestry. It is possible to upgrade the descendants of such individuals if their pedigree is found to be sufficiently 'pure'. British goats can appear in any colour or size depending on the breed they are associated with.

BREED ORIGINS

British goat breeds lying together.

Due to the diverse range of goats in this category, the British goat has become an important part of healthy goat breeding and improving the national stock.

PRODUCE

The British goat is kept mainly for milk, with some of the UK's highest yielding goats coming from this genetic mixture.

BRITISH ALPINE

A black goat with white Swiss markings, the British Alpine is an impressive animal in summer when its coat shines with a natural gloss. Known for their changeable temperaments, they are generally considered a breed for experienced keepers who relish a challenge. Their dark coats may lighten if there is a lack of copper in their diet, one of the most essential minerals to all goats for healthy development of the central nervous system, bone growth and hair pigmentation. They are an active breed and thrive on a free-range lifestyle rather than being kept indoors.

The British Alpine winner at the 62nd Annual Dairy Show at Olympia, London.

BREED ORIGINS

Developed in the early 1900s from breeds including the British primitive, Nubian and Toggenburg, which shares its Swiss markings.

PRODUCE

An excellent dairy goat, British Alpines are known for their long lactations, sometimes lasting up to two years. They also have good-sized teats for ease of milking and are able to keep producing through the winter.

BRITISH PRIMITIVE

OPPOSITE British primitive goat in Scottish woodland.

A relatively small goat, adult males grow to an average height of 60–70cm and weigh 45–55kg. They have long, thick coats that blend with their beard, and scimitar horns that twist and curve. Common colour patterns include white, tan, badger-face, grey, black, mahogany and light-belly.

BREED ORIGINS

Also referred to as the Cheviot, British Native goat, Old British goat or the British Landrace goat, this breed is thought to have existed in the British Isles since the end of the Stone Age. It was among the foundation stock for a number of modern standardised breeds, including the Anglo Nubian.

PRODUCE

The British primitive is now considered a rare breed, existing in isolated feral herds, nature reserves and some breeding stock on rare breed farms. Like the Bagot goat, they produce low milk yields and are best suited to conservation grazing purposes to diversify land for wildlife.

ENGLISH

Hardy, sturdy and docile, the coats of this breed has coats varying between brown, grey and black, with a characteristic dark line or 'eel stripe' running along the back. They usually also have dark markings on their head, legs and flanks, often with a few white patches. With a lovely temperament and content to be in human company, they will occasionally

English kids.

reveal the mischievous manner that makes them such fun to keep.

BREED ORIGINS

The first breeders' association was formed in the 1920s in an attempt to revive the smaller, hardier, lower yielding English goat. The present association was formed when goatkeepers discovered odd characteristics in stock from Lancashire, Dorset and Somerset, reminiscent of old breeds existing before others were imported. The association bred selectively from these animals in order to re-establish what is now known as the English goat.

English goat.

PRODUCE

The ideal breed for smallholders, English goats are great dual-purpose animals with a good conversion rate of milk and meat throughout the year. They are happy to eat a wide range of wild and prepared food, and are well suited to the changeable British climate and tough vegetation.

GOLDEN GUERNSEY

Easily identified by its beautiful golden coat, the colour can vary from pale blonde to deep bronze. The coat is usually medium to long with some fringing. The British Guernsey is a separate sub-breed, developed from the continual use of Golden Guernsey males on successive generations of female progeny. The British Guernsey is slightly larger than the Golden Guernsey, although they are not easily distinguishable.

BREED ORIGINS

A rare breed of goat from the Bailiwick of Guernsey on the Channel Islands, the Golden Guernsey breed was first imported to Britain by Rudi Sweg in 1965. The British sub-breed was

Golden Guernsey male.

developed in the years following. Theories suggest the Guernsey goat may share ancestry with the Oberhasli and Syrian breeds.

PRODUCE

An efficient milking goat considering its relatively small size, this breed can yield an average of 3kg per day with a high butterfat and protein content. Maiden milkers are also common, the term used to refer to nannies that come into milk without first having a kid.

PYGMY

Pygmy goats are genetically dwarfed animals kept for enjoyment and companionship. With short legs and cobby bodies, an adult male pygmy will reach a maximum height of around 56cm at the withers. Their coats can take any colour except completely white, but to meet show standards they are not allowed any Swiss markings. They have a generally quiet, docile temperament.

BREED ORIGINS

This breed was developed from the West African dwarf goat, found most commonly in the Cameroon valley. During

the colonial era they were imported into Europe and later acquired by private breeders, gaining popularity as pets.

Pygmy goats.

PRODUCE
The pygmy goat is the most common breed of goat kept as pets. They are too small for commercial meat production and milk dries up shortly after the kid stops suckling. They have good-natured, friendly personalities, but still need the space, care and attention of other domesticated breeds, so ownership should be considered carefully.

SAANEN
A distinctive white goat with a short coat, supple skin and a 'feminine' head. The British sub-breed has slightly longer legs and a heavier build, while the ears are erect and point upwards and forwards. Their natures are calm and pleasant, but they do not tolerate strong sunlight due to their pale skin.

BREED ORIGINS
The name originates from the Saanen valley in Switzerland, where selective breeding of dairy goats has taken place for

Saanen kids.

many centuries. The British Saanen sub-breed began with the introduction of goats from Holland in 1922, and a black variant known as the Sable Saanen is now a recognised breed in New Zealand.

PRODUCE

The Saanen is a popular breed for milk production throughout the year, due to their long lactation periods and high yields of milk with around 3.2 per cent fat and 2.7 per cent protein. Large groups are often housed together.

TOGGENBURG

The Toggenburg ranges from mid-brown to grey or fawn shades, with a silky coat of varying lengths and some degree of fringing. The face, lower legs and tail are usually decorated with Swiss markings, and tassels or toggles may be present – these are hollow pieces of gristle on their throats, covered with hair but with no known purpose. The British sub-breed is slightly larger and more often brown than grey, although lighter and darker colours

OPPOSITE
A Saanen doe in Wales.

Toggenburg
goats.

are acceptable. They are a strong, robust breed with a reputation for longevity.

BREED ORIGINS

The name derives from the Toggenburg region of the Canton of St Gallen where the breed is thought to have originated. It is now among the most productive breeds of dairy goat, distributed in at least fifty countries and on every continent.

Toggenburg
headshot.

PRODUCE

The Toggenburg is a highly productive dairy breed, with an average milk yield of 3.7 per cent butterfat and 2.71 per cent protein. The British sub-breed is one of the most popular goats in Britain and is often used on goat farms specialising in cheese production.

GOAT PRODUCE

WHILE A GROWING number of animals are raised for meat, the majority of the British goatkeeping industry is dedicated to the production of milk, cheese, yoghurt and butter. In 2005, a report in the *Telegraph* claimed that demand for goats' dairy products had increased so much that farmers were struggling to keep up. One possible reason for this is that goats' milk can often be tolerated by those who are allergic to cows' milk, as it contains fewer allergenic proteins. Some dieticians also claim the fat is more digestible, lactose and cholesterol levels are lower, calcium and potassium are high, and it contains more oligosaccharides, which act as a prebiotic in the gut.

Milking by hand is only as difficult as the goat you are working with. Quite understandably, some do not enjoy being forced onto the milking stand and milked, but there are few who can't be persuaded by a good bucket of food or a gentle brush of the coat to ease them along. The easiest way to learn is to contact your local breeding society and ask somebody to teach you. Remember the importance of keeping the entire milking process clean and hygienic, especially if you choose not to pasteurise your milk.

Pasteurisation is a process used by commercial dairies to extend the shelf life of a product.

The author milking one of her English goats.

Pasteurising milk.

It's a useful way to destroy harmful bacteria that could produce disease or cause spoilage, without radically altering flavour or quality. On the other hand, removing certain bacteria from milk used to make homemade cheese can cause the cheese to rot rather than mature, and some claim the natural bacteria is healthy for the body. A popular fermented drink known as kefir is made with fresh pasteurised goats' milk and live unpasteurised kefir grains, using the kefir bacteria to promote probiotic health benefits. Either way, pasteurisation has its own benefits and downsides. To pasteurise milk at home, warm it slowly to 30°C – taking care not to let it burn – and then remove it from the heat.

Cheesemaking, also known as caseiculture, is one of mankind's oldest industries. Some theories suggest the first cheese may have been created accidentally when milk was transported in bladders made of ruminants' stomachs. These stomachs contained a natural supply of rennet, a set of enzymes that separate milk into curds and whey to make cheese. Making your own goats' cheese is a great way to learn about the process and experiment with taste and texture, from the light, lemony notes of soft, crumbly cheese, to the rich, pungent aroma of blue moulds.

Separating curds and whey.

Compared with most other red meats, studies suggest goat meat, also known as chevon, is lower in calories, saturated fat and cholesterol, as well as being high in iron, protein and B vitamins. While it isn't as widely available in Britain as lamb and beef, the goat meat market is growing, possibly

due to claims it is a more sustainable alternative in the face of climate change. For British goatkeepers, there are also strict government regulations on how you raise, slaughter, butcher and distribute your meat. The flavour of goat meat has been compared with that of lamb, albeit milder and more tender, and just like lamb, the texture toughens as the animal ages. For those wanting to dabble, try adding goat meat to curry with the fresh flavours of ginger, thyme and lemon.

ABOVE LEFT Soft goats' cheese with chives.

ABOVE RIGHT Bowls of goat curry.

Goats' milk is rich in essential fatty acids and triglycerides, which are not only great at moisturising the skin, but also have a similar pH to humans, causing less irritation and absorbing easily into the skin. It also contains skin minerals like age-defying selenium, vitamin A, and lactic acid, which helps to gently remove dead cells and brighten the skin. Try making your own creamy goats' milk soap by combining a soap base with fresh milk and organic rose oil, or use lavender, coconut or apricot kernel oils for a range of natural aromas.

Homemade goats' milk soap.

Mohair is the name given to the wool of the Angora goat, whose smooth, silky coats are used to make luxurious garments and textiles around the world. Spinning animal fibres by hand is a craft that dates back thousands of years in Britain, and mohair is one of the easiest fibres for beginner spinners to work with. Thanks to its long staples, mohair can be spun with less tension than sheep's wool, although its slippery texture can take some getting used to. By the end, your yarn will shine with natural lustre, and can be used to knit jumpers, scarves or any other cosy item.

Baby mohair from kids is the finest and softest goat fibre, but also the most difficult to find. It can be blended with fine wool like merino, silk or Angora rabbit to create a high quality yarn skein. Adult mohair is slightly coarser but more durable, as well as costing less, and can be used to make a rug or thick socks.

Fleeces should be washed gently to remove grease and dirt without damaging the fine texture of the wool. If you decide to dye your wool, bear in mind that mohair absorbs dye more efficiently than sheep's wool, but it also requires a lower working temperature to avoid unnecessary damage. The fleece can also be carded using paddles or a machine, which removes uneven lumps and helps simplify the spinning process to create a smooth, soft yarn. When spinning by hand, remember that a looser twist in the yarn will bring out more shine but reduce durability, whereas a tighter twist will strengthen the yarn but decrease its natural lustre.

Spinning wool with a drop spindle.

Goats are increasingly used as grazing livestock for conservation, helping to manage delicate habitats for wildlife by eating competitive plants. Being excellent browsers, they can get all their nutritional needs from scrub, although as they are not particularly fussy this can lead to overgrazing if not properly managed. They are also highly mobile, being able to jump up to high places that even sheep might not reach, which means that

Basket of mohair wool.

landowners must keep fences in place to prevent their living lawnmowers from escaping.

Cheddar Gorge in Somerset is home to around sixty feral goats and a hundred feral Soay sheep. The goats originated from a mixture of older breeds, but they bear a strong resemblance to modern Bagot goats, which are a rare breed in Britain today. There are now around forty populations of feral goats across the UK, and the Cheddar flock was introduced in the 1990s to help keep its precious wildlife habitats in peak condition. The goats form part of a much larger ecosystem, and their daily routine of chomping away at competitive plants means there is more room for rarer wildflowers like the Cheddar pink, harebells, lesser meadow-rue, meadow saffron, herb-paris and dusky crane's-bill. A more diverse range of plants means more invertebrates and other wildlife exist here, too, such as peregrine falcons, ravens, and both the greater and lesser horseshoe bats that roost in the famous Cheddar caves.

A peregrine falcon, one of the beneficiaries of conservation grazing.

THE FUTURE OF GOATKEEPING

WITH AN ANNUAL population growth rate of 0.6 per cent in Britain, it's no wonder the demand for goat dairy products is almost outstripping the supply. It isn't just the increasing number of British citizens, but the fashion for goats' cheese and goats' milk that has seen a quiet revolution in the goat industry over the last few years. Today, there are around 450,000 goats in Britain, the majority of which produce milk for drinking and cheesemaking. These products are now on sale in every major supermarket, farm shop and high-end restaurant, the goat's clouded past as the 'poor man's cow' quite forgotten. But is goats' cheese and meat just a passing trend, or will the industry help shape the future of sustainable British farming?

According to the most recent estimates, meat and dairy production is now responsible for a fifth of global greenhouse gas emissions, and unless there is a shift in global habits this figure is predicted to double by 2050. At the same time, scientists claim that emissions must be cut by at least 80 per cent within the same period to prevent the worst effects of climate change. These effects include flooding, droughts, ocean acidification, mass wildlife extinction, food shortages and extreme weather. In the face of these statistics, many people have been empowered to make a positive change and switch to a vegetarian or vegan diet, reducing their own meat and dairy intake to reduce carbon emissions. In the UK alone there are now an estimated 3.5 million vegans and vegetarians, and the numbers are growing.

OPPOSITE
A beautiful feral goat.

Industrial
cheesemaking.

But for those who choose to eat meat and dairy, is there a way to source it more ethically? One argument claims that goat meat and milk is a more ethical source of nutrition than cattle and sheep, due to the way their bodies produce methane. A recent study in the French scientific journal *INRA* measured the methane emissions produced per kilogram of milk in sheep, goats and cows, and found that goats were one of the lowest emitters per yield. Goats' ability to forage on low-nutrient and waste ground is another reason they are heralded as a source of ethical meat, especially when they are used to manage nature reserves at the same time.

The issue is not, however, a simple one. Goats' milk is highly sought after in Britain, with many people happy to pay extra for delicious cheeses or a 'healthier' alternative to cow's milk. But in order to generate enough milk to meet demand, farmers are forced to breed a substantial number of female goats while slaughtering most males at birth. Not only are males unable to produce milk, but young kids will naturally drink the milk produced by their mothers. In contrast to the cattle industry, where young male calves are raised for veal or beef, the British demand for goat meat is no way near as high as our love for goats' milk. As a result, an estimated 90,000 male kids are slaughtered every year and disposed of as waste, a tragic statistic in a world where a third of global food is already wasted and 13 per cent of the world's population is undernourished.

Goats and
cattle sharing a
paddock.

So, what is the solution? Although the goat meat industry is still much smaller than other livestock industries, interest from the general public is growing. As people become more experimental with their diets and more dedicated to sourcing ethical food, goat meat is becoming a more normalised option, and if otherwise wasted goat meat could replace beef and lamb as a more ethical protein source, the balance might begin to right itself.

Traditional goatkeeping in Kashmir, India.

While we may not naturally think of it as traditional British cuisine, goat meat is extremely popular around the world and makes up 60 per cent of red meat eaten globally. In India, which is home to more vegetarians than the rest of the world combined, goats are more populous than sheep, and unlike cattle are not taboo to eat under dharmic law. Here, goat meat is often labelled as mutton, regardless of it being sheep or goat, and most street markets sizzle with spicy masala goat. Estimates suggest that huge proportions of the world's goat population are distributed across Asia, Africa and the Caribbean, with most goatkeeping communities simply sustaining themselves rather than expanding into a larger industry. Due to their hardiness, high yields and low nutritional needs, goats are a useful animal for those in developing nations, providing them with both meat and dairy with minimal maintenance.

In 2011, a census showed that around 13 per cent of the UK population was born overseas. Just seven years earlier, a research study by Ubamarket showed that, while we may cherish classic British dishes like roast dinner or fish and chips, in reality curry comes top of the list of our favourite foods. Spaghetti bolognaise comes second, followed by other imported dishes like lasagna and chilli. Never has the British menu looked more intriguing, and with the addition of flavours from around the world, our national cuisine has become a delicious hotpot of old and new. It comes as no surprise, then, that goat meat is on the rise. Not only are consumers more interested in the ethics behind their food, they have also been enticed by flavours and recipes from faraway lands. Chefs have started to champion goat meat as a delicious, sustainable ingredient, and while it still costs around the same price as organic lamb, goat meat can now be found in many high street butchers, farm shops and supermarkets.

It isn't just an agricultural revolution that has caused goat numbers to rise, but a cultural shift in Western trends. Goats are not only useful and resilient – they are now celebrated as cute, friendly and entertaining animals. Products like goat ice cream, fudge and milkshakes have started popping up in hip cafés, and YouTube is bursting with videos of pygmy kids, screaming goats and billies climbing trees. In the last few years, goat yoga or 'goga' has even emerged as a novelty take on the traditional Hindu practice. Born in America but quickly taking root across Europe, goat yoga involves a simple yoga session shared with a herd of goats who clamber over participants, nibble hair and generally cause mischief. It sounds silly, but this kind of animal therapy has been proven to decrease blood pressure, lower cholesterol and reduce feelings of depression and loneliness.

Whether you're a large-scale farmer, back garden smallholder, land manager or animal lover, the golden era of goatkeeping has returned to Britain and goats are once more in their prime. No longer a burden of the lower classes, an enemy of landowners, or an animal of devilish tendencies, goats in Britain are finally being recognised for what the rest of the world has always known they are – resilient and clever animals with a mischievous charm and loveable nature.

Goat yoga.

CRAFT AND RECIPE IDEAS

GOATS' MILK AND ROSE PETAL SOAP

When it comes to skincare goats' milk has lots of nourishing properties. This recipe uses the relaxing aroma of rose oil to create a soft, creamy soap.

INGREDIENTS

Silicone soap mould
450g 'melt and pour' soap base
50ml goats' milk
2 tsp rose oil
Handful of dried rose petals (optional)

METHOD

1. Cut the soap base into 2cm cubes with a clean knife.
2. Melt the cubes in a microwave or bain-marie until it forms a smooth liquid without lumps or bumps.
3. Allow the soap to cool to around 50°C, then add in the goats' milk, rose oil and petals, if using.
4. Stir well and store in the fridge until ready to use.

JAMAICAN GOAT CURRY

Like mutton, goat requires a long cooking time to ensure the meat is tender and absorbs the surrounding flavours. Serve with rice and peas for a taste of the Caribbean.

INGREDIENTS

Olive oil
750g goat shoulder fillet
1 lime
2 tbsp hot curry powder
2 onions
4 garlic cloves
3cm fresh ginger
3 red chillies
1 tsp dried sage
1 cinnamon stick
2 tsp ground allspice
400g tin chopped tomatoes
1 tbsp sugar
500ml vegetable stock

METHOD

1. Slice the shoulder fillet into 2.5cm pieces and place in a bowl with the juice and zest of the lime, 1 tbsp curry powder and a glug of olive oil. Combine and leave aside to infuse.
2. Peel and chop the onion, garlic, ginger and chillies into small chunks. Heat a splash of olive oil in a large frying pan and fry the chopped ingredients for a few minutes until they start to brown. Stir in the remaining curry powder, spices and herbs and cook for a further minute, before adding the goat slices and frying until sealed on all sides.
3. Add tomatoes, sugar and vegetable stock to the pan, reduce heat, cover and simmer for 1.5 hours before serving.

CHOCOLATE MILKSHAKE

An indulgent way to try goats' milk or use up any leftovers – don't forget the whipped cream!

INGREDIENTS

300ml pasteurised goats' milk
3 scoops chocolate ice cream
1 tbsp chocolate hazelnut spread
6 marshmallows

METHOD

1. Combine all ingredients in a blender and whizz until smooth and thick.
2. Serve in a tall glass with whipped cream, grated chocolate and an extra marshmallow.

STRAWBERRY AND THYME ICE CREAM

The perfect treat for a warm summer's day, the delicate flavour of this strawberry ice cream is lifted with a sprinkle of dried thyme.

INGREDIENTS

60g caster sugar
300g strawberries, rinsed
½ teaspoon thyme, rinsed and finely chopped
Zest of ½ a lemon, grated
150ml goats' milk
150ml double cream
3 egg yolks, beaten

METHOD

1. Pop the strawberries, chopped thyme and lemon zest into a bowl and gently mash with a fork to release the juices. Sprinkle with 30g of the caster sugar and leave covered for an hour.

2. Add the milk and cream to a saucepan and bring to boil over a medium heat.

3. In a separate bowl, combine the remaining sugar and beaten egg yolks until mixed thoroughly.

4. Pour the warm milk mixture over the eggs and whisk well. Then return the whole mixture to the pan and stir continuously over a medium heat until it thickens into custard.

5. Fold in the strawberries, thyme and lemon and beat well. If you prefer your ice cream without chunks, you can now add the mixture to a processor and blend until smooth. When you are happy with the consistency, pour the mixture into a plastic tub and store in the freezer until solid.

FURTHER READING

Amundson, Carol. *How to Raise Goats: Everything You Need to Know*. Voyageur Press, 2019.

Caldwell, Gianaclis. *Holistic Goat Care*. Chelsea Green, 2017.

Crowther-Smith, Alison. *Silky Little Knits: Luxurious Designs and Accessories in Mohair-Silk Yarns*. Trafalgar Square, 2009.

Dunbarn, Edward. *Pygmy Goats as Pets*. Zoodoo, 2018.

Ekarius, Carol, and Robson, Deborah. *Fleece and Fiber Sourcebook*. Storey, 2011.

Faiola, Anne-Marie. *Milk Soaps*. Storey, 2019.

Fleming, Anne. *The Goat*. Pushkin, 2018.

Hitch, Valerie. *Goat Keeping: The Art of Good Husbandry*. Interpet, 2009.

Horan, Kevin and Elena Passarello. *Goats and Sheep: A Portrait Farm*. Five Continents, 2019.

Hurst, Janet. *The Whole Goat Handbook: Recipes, Cheese, Soap, Crafts & More*. Voyageur Press, 2013.

Kessler, Brad. *Goat Song: A Seasonal Life, A Short History of Herding, and the Art of Making Cheese*. Scribner, 2010.

Lauricella, Leanne. *Peace, Love, Goats of Anarchy*. Rock Point, 2018.

Matthews, John G. *Diseases of the Goat*. Wiley-Blackwell, 2016.

McGaha, Jennifer. *Flat Broke with Two Goats: A Memoir*. Sourcebooks, 2018.

McLaughlin, Chris. *Raising Animals for Fiber*. Companion House, 2019.

Morse, Lainey. *The Little Book of Goat Yoga: Find Your Farmyard Flow*. Yellow Kite, 2018.

Nix Jones, Shann. *Secrets from Chuckling Goat*. Hay House, 2015.

Nix Jones, Shann. *The Kefir Solution*. Hay House, 2018.

Whetlor, James. *Goat: Cooking and Eating*. Quadrille, 2018.

Zimmerman, Brent. *Get Your Goat: How to Keep Happy, Healthy Goats in Your Backyard, Wherever You Live*. Quarry, 2012.

PLACES TO VISIT

Butser Ancient Farm, Chalton Lane, Waterlooville,
Hampshire PO8 0BG. Telephone: 02392 598838.
Website: www.butserancientfarm.co.uk – experimental
archaeology and rare breed site.

Mucky Bucket Farm, 12 Woodthorpe Gardens, Sarisbury,
Southampton SO31 7AR. Telephone: 07527 256861.
Website: www.muckybucketfarm.co.uk – offers
goat yoga classes.

Cheddar Gorge, Cheddar, Somerset BS27 3QF.
Telephone: 01934 742343. Website:
www.cheddargorge.co.uk – land managed by National Trust.

Snowdonia National Park, Gwynedd, north-west
Wales. Telephone: 01286 679686. Website:
www.visitsnowdonia.info – great for spotting feral goats.

Museum of Islay Life, Port Charlotte, Isle of Islay, Inner
Hebrides PA48 7UA. Telephone: 01496 850358.
Website: www.islaymuseum.org – island is home to
primitive goats.

Buttercups Sanctuary for Goats, Wierton Road, Boughton
Monchelsea, Kent ME17 4JU. Telephone: 01622 746420.
Website: www.buttercups.org.uk – rescue centre for
abandoned, neglected and abused goats.

Rare Breed Goats, Kingdom Farm, Errol, Perthshire,
Scotland PH2 7SW. Email: richard@rarebreedgoats.co.uk.
Website: www.rarebreedgoats.co.uk – rare and traditional
breed goat conservation farm.

Museum of English Rural Life, 6 Redlands Road, Reading
RG1 5EX. Telephone: 0118 378 8660. Website:
www.merl.reading.ac.uk – part of the
University of Reading.

Northumberland National Park, Hexham, Northumberland
NE46 1BS. Telephone: 01434 605555. Website:
www.northumberlandnationalpark.org.uk – home to
feral Cheviot goats.

The author with her goats at the Singleton Show, Sussex.

Museum of Witchcraft and Magic, The Harbour, Boscastle, Cornwall PL35 0HD. Telephone: 01840 250111. Website: www.museumofwitchcraftandmagic.co.uk – good for evil goats.

Trowbridge Museum, Court Street, Trowbridge, Wiltshire BA14 8AT. Telephone: 01225 751339. Website: www.trowbridgemuseum.co.uk – specialists in the history of the wool trade.

Mudchute City Farm, Pier Street, Isle of Dogs, London E14 3HP. Telephone: 020 3670 7515. Website: www.mudchute.org – rare breed farm in the heart of east London.

INDEX